Microbiomes and Microbial Communities 2017

Topical Conference at the 2017 AIChE Annual Meeting

Minneapolis, Minnesota, USA
29 October - 3 November 2017

ISBN: 978-1-5108-5782-7

Printed from e-media with permission by:

Curran Associates, Inc.
57 Morehouse Lane
Red Hook, NY 12571

proceedings.com

Some format issues inherent in the e-media version may also appear in this print version.

Printed by Curran Associates, Inc. (2018)

For permission requests, please contact AIChE
at the address below.

AIChE
120 Wall Street, FL 23
New York, NY 10005-4020

Phone: (800) 242-4363
Fax: (203) 775-5177

www.aiche.org

Additional copies of this publication are available from:

Curran Associates, Inc.
57 Morehouse Lane
Red Hook, NY 12571 USA
Phone: 845-758-0400
Fax: 845-758-2633
Email: curran@proceedings.com
Web: www.proceedings.com

TABLE OF CONTENTS

Some abstracts in this book refer to images that were not made available by the publisher

510408 Using Metabolomics and 16S rRNA Sequencing to Investigate the Impact of Environmental Chemical Perturbations on Gut Microbiota Community Composition and Function

Abstract:

Recent findings suggest that a significant alteration of the gut microbiota, or dysbiosis, contributes to various diseases and disorders, including developmental disorders of the brain. Dysbiosis may result from infection, diet, or other environmental perturbations. One example of an environmental perturbation is exposure to biologically active synthetic chemicals present in household and commercial products. In recent years, exposure to these chemicals, labeled as endocrine disrupting compounds, have emerged as a significant public health concern. Interestingly, several of these chemicals, e.g. bisphenol A and phthalate esters, have been linked to neurodevelopmental disorders, including autism spectrum disorder (ASD).

This presentation describes a study on the effects of di-ethylhexyl phthalate (DEHP), a ubiquitous plasticizer. The effects of this chemical were studied in an *in vitro* culture model of the gut microbiota. Using 16S rRNA sequencing and untargeted LC-MS based metabolomics, we found that DEHP significantly modifies both the microbiota community structure as well as metabolic profile. Applying a novel metabolomics data annotation tool developed in our laboratory, we confidently identified more than 100 products of microbiota metabolism. Co-analyzing the 16S and metabolomics data using a metabolic model of the microbiota revealed that the chemically induced increases and decreases in specific metabolites can be attributed to the depletion or enrichment of particular groups of bacteria. Notably, DEHP exposure significantly increased the level of *p*-cresol, while expanding the abundance of *Clostridium bolteae*, both of which have been identified recently as potential biomarkers of ASD. Our results suggest that environmental chemicals could cause significant dysbiosis of the gut microbiota leading to an altered milieu of bioactive metabolites in the intestine, consistent with other studies linking environmental chemical exposure to developmental disorders involving gastrointestinal conditions.

510409 Genome-Scale Engineering of Lactococcus Lactis for Probiotic Outcomes

Abstract:

Humans have long reaped the benefits of probiotics via food fermentation with lactic acid bacteria. In this talk, I will discuss efforts in our research group to engineer the lactic acid bacterium Lactococcus lactis for improved probiotic outcomes. For example, besides its use in fermentation and in industry as a lactic acid producer, L. lactis has also been engineered for delivery of therapeutic proteins in vivo in the gut environment. We seek to use the toolbox of synthetic biology to expand the capabilities of this organism as an engineered probiotic. Our approach takes two routes: (1) Increasing the efficiency of genome engineering using the versatile CRISPR-Cas9 technology in this organism and (2) Discovery of genes and pathways that contribute to complex phenotypes by creation of genome-scale libraries and selection for desired phenotypes such as chemical tolerance and protein secretion. We also discuss the use of prebiotics to synergistically improve outcomes by encouraging lactococcal growth in complex environments. Increased throughput of genome engineering and identification of genomic targets will broaden the range of uses of this already food-safe organism for both industrial and human health purposes.

510410 Harnessing the Power of the Human Gut Microbiome to Treat Debilitating Disease

Abstract:

The human gut microbiome has become a hot topic in the search for new therapeutics to solve unmet medical needs. As the world's knowledge has expanded, the gut microbiome has been implicated in many conditions as far ranging as infectious disease to autism. In 2011, Rebiotix was founded to develop a microbiome-based drug to treat recurrent *Clostridium difficile* infection – a highly contagious disease of the gastrointestinal track – where standard of care therapeutics fall short. Join Lee Jones, UMN ChE '79, as she shares the journey and the science behind what is expected to be the first commercially available microbiome product in the world.

493110 Probiotic *Escherichia coli* Inhibits Biofilm Formation of Pathogens

Abstract:

Numerous chronic bacterial infections are involved in biofilm state and it is difficult to eradicate them by antibiotics. Biofilm formation is an inter-species phenomenon that involves dynamic interactions between bacteria in the biofilms. However, little is known to understand molecular interactions between biofilm bacteria especially in multi-species biofilms while majority biofilm studies has been done in mono-species. *Escherichia coli* strain Nissle 1917 (EcN) is a probiotic bacterium that has antagonistic effects on adherence, growth, and biofilm formation of other *E. coli* strains. In this study, we investigated the ability of EcN to outcompete with the biofilm formation of diverse pathogens such as enterohemorrhagic *E. coli* (EHEC), *Enterococcus faecalis*, *Pseudomonas aeruginosa*, and *Serratia marcescens*. When dual-species biofilms are formed, EcN reduced EHEC biofilm population up to 100-fold compared to EHEC mono-species biofilm formation, *E. faecalis* 10-fold, and *P. aeruginosa* 10-fold. EcN was dominant in the dual-species biofilms by occupying up to 90% of the total biofilm population. To see whether the molecules secreted from EcN affect EHEC biofilm formation, we grew EHEC biofilms under EcN cell-free supernant. EcN-supernatant inhibited EHEC biofilms up to 100-fold which is similar level of reduction to the result of EcN-EHEC biofilms. As a control, non-probiotic and non-pathogenic *E. coli* BW25113 did not reduce the biofilm formation of pathogenic bacteria. These results imply that probiotic *E. coli* inhibits the biofilm population of pathogenic bacteria, and provide insights that probiotics can be applied as an effective treatment for the biofilm-related infectious diseases.

498115 Understanding and Engineering the Metabolic Environment of a Healthy Skin Microbiome

Abstract:

The skin is the largest organ and the first line of protection against external factors. The skin environment is generally cool, acidic, and dry with inter- and intra-personal variation in pH, temperature, moisture content, elasticity and metabolite profiles. Different skin sites are colonized by diverse bacteria from *Actinobacteria*, *Firmicutes*, *Proteobacteria* and *Bacteroidetes* and fungi from *Ascomycota* and *Basidiomycota*. Some of these organisms are beneficial and contribute to protective properties of the skin. Skin commensal bacteria influence wound healing and infection by pathogens through direct antimicrobial activity and signaling host cells to boost innate and adaptive immune responses. To study these interactions and harness the protective power of commensal skin organisms, we are developing *in vitro* systems to monitor metabolic and immune responses of microbial skin isolates and skin cell cultures. Metabolic predictions of microbial growth states indicate that the skin pathogens are highly complementary with commensal isolates, suggesting an ecological mechanism for colonization. *Staphylococcus aureus*, a pathogenic skin organism, produces a small molecule which kills cultured skin models, while *Staphylococcus epidermidis,* a commensal isolate, protects cultured skin cells from pathogen growth. Further, in one-on-one interaction assays *S. epidermidis* inhibits the growth of *S. aureus* and other pathogens. Towards understanding the physical and chemical interactions important for the skin microbiome, we have developed reporter strains for skin isolates and have imaged infection dynamics using confocal and time-lapse microscopy. Collectively, our experiments suggest that commensal strains can have direct antimicrobial activity against pathogens and can activate host immune responses to further inhibit growth of pathogens. Improved understanding of these systems will allow us to examine interactions between resident microorganisms of the human skin and ultimately enable engineering the metabolic environment of the skin to encourage growth of beneficial microorganisms and enhance human health.

499557 Micro-Droplet Co-Cultivation and Characterization of Vaginal Bacteria

Abstract:

The vaginal microbiome (VMB) plays a fundamental role in womenâ€™s health and their susceptibility to diseases. For instance, Bacterial vaginosis (BV), the most common genital tract disorder in women of reproductive-age in the United States, is associated with preterm birth, an increased acquisition of HIV, and a higher risk of developing pelvic inflammatory disease. BV is characterized by the replacement of *Lactobacillus* species by other obligate and facultative anaerobic bacteria, such as *Gardnerella vaginalis.* Despite its significance, the fundamental questions of how diverse microorganisms in this microbial community interact with one another and with their host still remain largely unanswered. Current approaches for investigating these questions include conventional *in vitro* microbial cultivation and animal models, of which each has severe limitations such as low throughput and high resource requirement. More recently, microfluidic droplet co-cultivation has emerged as a new technology for dissecting microbial communities in a high-throughput fashion that renders substantial reduction of required reagents, time, and labor. In this work, we employ microfluidic droplet co-cultivation and several downstream analysis methods to recapitulate an antagonistic relationship between *G. vaginalis,* associated with BV, and *Lactobacillus jensenii,* a favored lactic acid producing vaginal bacterium. Pools of droplets can be characterized readily by fluorescence *in situ* hybridization (FISH) or quantitative polymerase chain reaction (qPCR). Furthermore, we show isolation of single droplets and colony formation on agar plates for quantitative analysis based on cell viability. Additionally, multiple displacement amplification (MDA) amplifies DNA content within a single droplet enabling metagenomic analysis. This work demonstrates the foundation of our microdroplet technology and represents the first step of our efforts aiming to elucidate microbial interactions in the human vagina and subsequently to intervene rationally the vaginal microbiome through engineering of probiotics and prebiotics.

501327 Targeted Species Removal in a Model Microbiome Via Phage Lytic Enzymes

Abstract:

Surveying microbiome composition (genomics, metagenomics) and assessing function through metatranscriptomic and metabolomic approaches have transformed our understanding of biodiversity and the roles microbes play in human health. However, new approaches and technologies are needed to develop mechanistic and causative insights into why and how communities vary over space and time, how communities adapt to change, and to develop new strategies for predictably modulating microbial community composition and function. Here, we focus on the targeted removal of a single species in a community. While adding an organism to an established model microbiome is relatively easy, the targeted removal of one species is more challenging. To address this challenge, we have developed a targeted bacteriophage lytic enzyme-based approach to selectively kill one organism in a model microbial community. Unlike conventional antibiotics, phage lytic enzymes have exquisite species specificity that target a single bacterial species or even a strain within a species. These lytic enzymes possess one of several cell wall peptidoglycan hydrolase activities with highly selective cell wall-binding specificities, which combine to confer activity and specificity toward a specific bacterial species. We have observed that *Staphylococcus aureus* can be selectively removed from co-cultures using the phage lytic enzyme lysostaphin (LST). Similarly, a second cell lytic enzyme, PlyPH, can be used to selectively remove *Bacillus cereus*. We have observed that the effectiveness of the selective killing depends on population density, lytic enzyme concentration and growth conditions. We will also present our work towards using Lst and PlyPH to selectively remove these organisms from a model human skin microbiome. *S. aureus* is a skin commensal in more than 30% of humans, while *B. cereus* is an opportunistic human skin pathogen. Such efforts are key to our long-term goal of being able to predictably modulate a target microbiome, based on causative or mechanistic insights, in order to alter microbiome function, or improve human health. The proposed project will also lay the groundwork for using targeted antimicrobial strategies for exploring microbiome dynamics and function in broad range of communities and environments.

504512 Metabolic Modeling of Microbial Communities

Abstract:

Despite rapid progress, our ability to probe microbial communities for species-specific information of metabolite pools or metabolic fluxes remains limited. Most of the available information refers to species abundance and overall chemical exchanges with the environment. Computational modeling provides the missing link for explaining the metabolic interactions of microbial partners in communities in the context of conservation laws and optimality principles. While several modeling techniques have been developed for microbial communities, little emphasis has been placed on the need to impose a time-averaged constant growth rate across all members for a community to ensure co-existence and stability. To this end, we developed the SteadyCom optimization framework for predicting metabolic flux distributions consistent with the steady-state requirement. SteadyCom imposes restrictions on the allowable community species membership, composition and phenotypes. In contrast to the direct use of FBA, SteadyCom is able to reveal hidden interactions in a community as demonstrated in a community of four *E. coli* auxotrophic mutants and predict an abundance profile of a gut microbiota model with a good agreement to experimental data.

In separate efforts, we applied community metabolic modeling to analyze experimental observations in the gut microbiome. A working hypothesis regarding the production of a beneficial estrogen, (S)-equol by gut microbes was assessed in terms of metabolic consequences by simulating the interactions between three representative organisms (Bacteroidetes, Firmicutes and equol producer). The effect of diet and species abundance on the communityâ€™s ability to produce (S)-equol was explored and testable predictions were generated. In another case study, a ten-species gut microbiota metabolic model was developed to analyze the effect of bile salt hydrolase (BSH) activity and gut microbial productions under the treatment by an obesity-related drug, glycine-Î²-muricholic acid. Under the running hypothesis of the studies was that BSH activity expressed in *Lactobacillus* and Clostridia relieves the intestinal farnesoid X receptor (FXR) antagonism and in turn induces obesity, short-chain fatty acid productions and amino acid consumptions by the gut microbiota correlated to the experimental changes was predicted, supporting the hypothesis. These case studies allude to the promise of genome-scale metabolic modeling approach in microbiome research.

504513 Micro-Droplet Enabled Co-Cultivation, Isolation, Andinteraction Characterization of Microbial Communities

Abstract:

Microorganisms live in synergistic communities in nature and play key roles in numerous natural or engineered systems. For example, there are increasing evidences that microbes in our gut assist in digesting food, help prevent the invasion of pathogens, and stimulate the immune system. However, the basic physiological and ecological mechanisms governing these complex microbial communities remain largely unclear. We have been exploiting droplet microfluidics for co-cultivation and dissection of symbiotic microbial communities. By encapsulating subsets of a microbial community in nano-liter droplets, we provide highly parallel localized environments for co-growth and aim to identify microbial interactions. The conceptual and technological framework is being applied to the investigation of the human gut and vaginal microbiomes.

510413 A Synthetic Probiotic Engineered to Treat Phenylketonuria

Abstract:

Phenylketonuria (PKU) patients are unable to properly metabolize the amino acid phenylalanine (Phe), leading to neurotoxicity. Currently, few broadly-useful treatment options exist. To address this unmet need, we constructed a synthetic probiotic strain of the *Escherichia coli* Nissle, designated SYN-PKU. SYN-PKU is engineered to express Phe degradative genes that are transcriptionally activated in the anoxic conditions of the mammalian gut. In phenylketonuric mice, SYN-PKU catalyzed a significant decrease in blood Phe levels independent of dietary protein intake, suggesting active recirculation of systemic Phe to the intestinal tract. Phe enterorecirculation was also demonstrated for the first time in healthy non-human primates (NHPs), and recirculating Phe was converted by SYN-PKU to *trans*-cinnamate, which was hepatically metabolized to hippurate and excreted in the urine, allowing its use as a quantitative biomarker of probiotic activity. Importantly, SYN-PKU also inhibited serum Phe elevation in NHPs following an oral Phe dietary challenge. This work supports synthetic probiotics as a novel class of therapeutics for treating rare genetic metabolic disorders and defines a strategy for their progression.

498718 *In vitro* Fermentation to Understand Healthy and Stressed Gut Microbiome Metabolism

Abstract:

The abundant bacterial consortia within the gut microbiome plays a critical role in metabolism of dietary compounds that translate to beneficial influences on the host. From a military perspective, the gut microbiome serves as an ideal tool to not only enhance Soldier gut and immune health, but also improve survivability and performance. Our work employs *in vitro* fermentation as a means to elucidate the systematic processes of colonic bacterial metabolism of dietary inputs under both native and stressed conditions. The broad experimental capacity of *in vitro* fermentation enables the generation of multi-faceted data sets (i.e time- and dose-dependent, colonic domain specific, etc.) to gain a detailed understanding on bacterial community dynamics, inflammation and immunity. This talk will focus on the use of *in vitro* fermentation to investigate both the prebiotic potential of cranberry A-type proanthocyanidins (PAC) within a healthy microbiome, and the utilization of fermentable fiber and protein as a function of stress-induced dysbiosis on gut metabolism. Briefly, fecal inocula derived from three individuals was used to inoculate a nutrient-rich anaerobic media supplemented with purified PAC. Colonic fermentations were performed in parallel at 37°C under colonic domain-specific condtions using an automated bioreactor platform for precise control over a multitude of parameters. Total growth assessment through protein content analyses revealed domain-specific metabolism. After 5 hrs, PAC caused a 2.3-fold increase in total bacterial abundance under transverse colon conditions while ascending and distal colon showed 1.7 and 1.3 fold increase respectively. After 10 hrs, further PAC metabolism was not evident in ascending and distal colons; however, was still evident under transverse conditions until stationary phase was met. Bacterial population dynamics, determined through 16S rRNA sequencing, metabolic byproduct analysis via GC-FID and PAC metabolite discovery through LC/MS is underway. These analyses will enable identification of specific bacteria and the upregulation of certain SCFAs that collectively result in a beneficial health effect on the host. For stressed metabolism studies, fecal samples from volunteers exposed to an acute military-relevant stressor, a sudden change in diet for 21 consecutive days, were pooled (N=10) and inoculated in a nutrient-rich medium supplemented with resistant starch (RS II) or casein, at multiple dosages. Total growth, population and SCFA analysis for comparison to the pre-intervention microbiome state and to volunteers on a standard diet (N=10) will be discussed. Efforts to simulate the gut microbiome using continuous culture fermentation will also be presented. An understanding of gut microbiota metabolism dynamics under both stressed and unstressed conditions could direct future dietary supplementation strategies to build resiliency against military-relevant stressors and offset negative health and performance impacts of gut dysbiosis.

500987 Phylogenetic and Functional Profiling of the Microbiome Associated with a Phytoremediation System Targeting Air Quality in the Built Environment

Abstract:

We are spending majority of our time indoors, where we can be exposed to harmful chemicals and bacteria. The active modular phytoremediation system (AMPS) is a biowall where plants are grown aeroponically in a granular bulk media. Biowalls have strong potential for improving indoor air quality through bioremediation by both plants and their-associated microbes. Understanding whether the presence of soil affects the AMPS microbiome was targeted because the presence of soil in biowalls has been correlated to the presence of mold spores and increases in plant module replacement. We wanted to know if microbial community composition and function are maintained if soil is removed from the AMPS modules and plants are supported only by the granular bulk media. Microbiome samples were isolated from bulk media and roots from three modules without soil and bulk media, soil and roots from three modules containing both soil and bulk media. We determined the phylogenetic profile by sequencing the 16S rRNA gene. All microbiome samples were observed to contain diverse communities with many rare species, and were all dominated by members of the phyla *Proteobacteria* and *Planctomycetes*. These phyla are typical of plant-associated microbiomes. Small differences are observed in the number and diversity of species in roots from modules with and without soil. Samples were clustered based on abundant and rare community members by calculating pairwise Bray-Curtis and Canberra distances respectively. In both cases, root and soil samples from modules containing soil clustered separately from all other samples. Thus, the presence of soil does contribute to a minor increase in diversity and drives differences in community composition. We are exploring if genes capable of promoting plant growth or degrading pollutants such as volatile organics are present in the communities. These functions may be leveraged to enhance phytoremediation using AMPS-type systems. We are using a novel computational tool to query the presence of these target genes using the 16S rRNA data and available genomic and metabolic databases to make functional predictions. Predictions can be verified using PCR-based techniques. This study demonstrates that the microbiomes associated with plants grown with and without soil in the biowall are diverse. The results set the stage for forward engineering efforts to enhance the remediation capabilities of these microbiomes and improve indoor biodiversity.

488575 Localization of Matrix Production Reveals *B. subtilis* Biofilm Growth Mechanics

Abstract:

Bacterial biofilms are communities of microbial organisms that are encased in self-produced extracellular polymeric substances (EPS). The EPS matrix facilitates cellular signaling that leads to coordinated behavior at scales that are much larger than the single bacterium. This gives rise to sophisticated collective behavior that is absent in planktonic cells, as evidenced by the enhanced mechanical rigidity, advanced architecture for water retention and uptake of nutrients, and the elevated antibiotic resistance in biofilms. Existing studies have mainly focused on the regulatory networks involved in biofilm formation at the single-cell level. The relation between colony-scale physiological growth and behavior, with spatial and temporal patterns cell-di erentiation within the colony remains under-explored.

Here, using distinct fluorescent transcriptional reporters and *in situ* imaging of biofilm growth, I demonstrate that considering the spatial and temporal distribution of the matrix-producing, sporulating and motile cellular phenotypes is critical in resolving *B. subtilis* biofilm development. Our key result is that, beyond an initial transient phase, the production of the EPS matrix and the onset of sporulation is localized to a radially propagating front at the biofilm exterior. The width of the matrix producing front approaches a constant value at late times. By using a correlation analysis in the fluctuations of fluorescence reporter activity, we reveal that the propagating fronts correspond to a pair of travelling waves of *tapA* gene expression (matrix-related) and *sspB* gene expression (sporulation-related) of cells that are immobilized within the biofilm matrix. Importantly, we show that the spatiotemporal propagation of both traveling waves are coupled by a single length scale (~ 500 μm) and time scale (~1.4 h) throughout development. Furthermore, the spatial gene expression profiles of matrix production and sporulation exhibit a data collapse into a self-similar asymptotic shape, and the front displacement exhibits a square-root scaling - hallmarks of universal dynamics that governs matrix production and sporulation of cells within the front.

What are the implications of our discovery towards quantitatively characterizing physical mechanisms that give rise to the observed phenotypic diversity and physiological heterogeneity? In a single cell, it is well-known that nutritional stresses and starvation triggers a transition from matrix production to the transcription of genes involved in sporulation. Additionally, the physical expansion and spreading of the biofilm colony must result from the mechanical forces associated with EPS matrix production within the localized propagating front. Therefore, a full biomechanical model of biofilm shape and morphology requires coupling nutrient uptake and consumption that gives rise to localized patterns of matrix gene expression, with mechanical forces that govern colony expansion.

In the second part of the talk, I present a mathematical model that combines nutrient transport, a concentration dependent biomass production rate and a thin-film equation that balances biomass production and biofilm advection driven by mechanical forces due to osmotic pressure and cell growth. I show that the model reproduces both the shape and morphology of the biofilm colony, observed expansion rates, as well as the experimentally measured matrix-production and sporulation gene expression profiles. Finally, I discuss the broader implications of using this combined experimental and modeling approach to probe various aspects of understanding biofilm colony development, phenotypic diversity and growth strategies.

501420 Development and Characterization of a Promoter Library for Probiotic *E. coli* Nissle Using a High-Throughput Barcoding Approach

Abstract:

Knock-in of genes or pathways with therapeutic potential to the gut microbiota using engineered probiotic strains has the potential to revolutionize human health. However, the dosage of therapeutics delivered *in vivo* must be precisely controlled both spatially and temporally. This degree of tunability requires detailed knowledge of transcriptional regulator activity *in situ*. To date, engineered organisms have been largely developed in the aerobic, homogeneous, resource-rich, noncompetitive environment of the test tube. The gut environment is largely the opposite, so it is unclear how well existing databases of transcriptional regulators will function in gut-resident probiotics. However, accurate determination of gene expression in the gut context requires a vertebrate host, quickly making the traditional approach of testing each promoter in separate animals infeasible and unethical. We have optimized a high-throughput approach for *in situ* measurement of promoter activity using DNA barcodes, enabling expression assays for entire promoter libraries to be performed in single animals using high-throughput sequencing. To demonstrate this approach, we curated 32 candidate promoters through analysis of metatranscriptome datasets and promoter libraries, and transferred these synthetic constructs into *E. coli* Nissle. We confirmed that our barcoding strategy had an undetectable effect on gene expression, but interestingly found that each barcode exhibited a significant (up to 5-fold) and reproducible bias during amplification and sequencing. In response, we developed a model of this bias enabling accurate recapitulation of known ratios of barcode-containing DNA in standard curve experiments. Using this validated method, we then analyzed promoter activity in *E. coli* Nissle grown *in vitro* and *in situ* under a variety of environmental conditions (oxygen content, diet, and body site). Collectively, these experiments reveal exciting dynamics in synthetic constructs during passage through the mammalian gut and inform the design of probiotic strains engineered to improve human health. More broadly, these experiments also reveal PCR bias as a significant confounder of barcoded abundance measurements and suggest strategies to overcome its effects.

510417 Unlocking the Potential of the Animal Mycobiome for Bioenergy and Animal Health

Abstract:

Complex microbiomes within the digestive tract of herbivores are increasingly of interest for the discovery of novel lignocellulose degrading enzymes and biocatalysts for the production of biofuels. Central to this community are anaerobic fungi, which are prolific digesters of crude lignocellulose through a combination of powerful secreted enzymes and invasive growth. Despite their outsize contributions to digestion in large herbivores, the mycobiome has historically been overlooked due to its low abundance within the microbiome. Here, we enrich for these fungi and pursue a multi â€“omics approach integrated with biochemical characterization, growth phenotypes, and heterologous expression to probe the catalytic potential of these overlooked organisms for bioenergy and improved animal health. We have assembled a panel of novel fungal isolates from both zoo and farm animals representing several types of fungal genera. We demonstrate that these fungi express and secrete tightly regulated carbohydrate active enzymes (e.g. cellulases) that are active on a broad array of untreated agricultural residues, food wastes, and woody forestry products. More importantly, these enzymes are robust against increasing levels of syringyl lignin, which is a natural plant defense against microbial degradation. Thus, these anaerobic fungi have significant potential to pretreat diverse and difficult lignocellulosic feedstocks for bioenergy production without enzyme supplementation from other organisms. Also encoded within the genomes of these organisms are large reservoirs of natural product gene clusters that rival those of super-producing actinomycetes. These natural products are hypothesized to mediate microbiome function, and thus animal health, and offer potentially novel biocatalytic routes to advanced biofuels and medicines. We are currently validating these clusters and characterizing their products and regulation through metabolomics and co-culturing studies. Our ongoing efforts suggest that the animal mycobiome is a rich untapped reservoir for bioenergy, and provides a roadmap for its application in diverse engineering efforts.

510418 40,000+ Genomes and Counting: Computational Lessons from Building a Giant Culture Collection

Abstract:

Biological sequence data is as rich and complex as it is inexpensive to produce, and the pace of DNA sequencing technology development quickly renders current best practices and conventions for storing and organizing biological sequence information obsolete and/or irrelevant. This is a particular challenge for companies specializing in microbiome science and amassing large collections of microbial isolates -- how can we account for this ever-changing environment and maintain a stable data infrastructure at scale? At AgBiome we have selected a technology stack for genomic data science that enables rapid development, persistent relevance of analyses, and reproducibility -- we employ suite of opensource tools including NextFlow, Docker, Mash/Sourmash, and Jupyter. In this talk, I will contrast data analysis in academia versus industry and present an analysis case study involving one of AgBiome's products in development to demonstrate how we battle the head-spinning dynamism of genomics.

492809 Engineering Cross-Feeding Co-Cultures As a Platform for High-Throughput Screening of Microbial Strain Libraries for Enhanced Biomolecule Production

Abstract:

A variety of methods have been developed to explore vast phenotype landscapes in a controlled manner. The applicability of these methods to the development of bacterial production strains is, however, limited by the requirement of a screening method with high enough throughput to investigate all the variants generated. Products that require chromatography-based screens typically reach a throughput of only ~10^2 variants per day, substantially lower than those with colorimetric or growth-based assays. Here, we present a new method for screening *Escherichia coli* strains for improved production of amino acids and/or their intermediates. Amino acid intermediates can be diverted towards production of a variety of other useful compounds, including higher alcohol biofuels.

This method uses a cross-feeding auxotroph system to convert production levels of amino acid or amino acid intermediates into co-culture growth characteristics that are detectable in a high-throughput context. As proofs of concept we examine two target molecule cases: tryptophan and 2-ketoisovalerate (a precursor of valine, leucine, and the drop-in biofuel isobutanol). We demonstrate that suitable pairs can be selected so that improvements in production by the production strain lead to increased co-culture growth rate and changes in the final composition of the population. We explore implementation of this co-culture screening method on petri dishes and in microdroplets, demonstrating its ability to identify rare over-producers from a library. We investigate the dynamic ranges of our model systems and ways in which they may be expanded.

495657 Engineering Microbial Consortia for Bioelectrocatalysis Processes

Abstract:

Bioelectrochemical systems hold great promises for diverse bioenergy and environmental applications, including microbial fuel cells (MFC) for green and sustainable bioelectricity production from wastewater and biomass, autonomous power sources for sensors and beacons by marine sediment MFCs, sea waters desalination by microbial desalination cells, hydrogen production by microbial electrolysis cells, and microbial electrosynthesis for CO_2 reduction and chemicals production.

However, exoelectrogens, such as *Shewanella* and *Geobacter* being widely studied in MFCs, could only use limited spectrum of carbon sources. To expand the carbon source range being used in MFCs, we herein rationally designed a few microbial consortia to broaden the carbon sources that could be used in MFCs and bioelectrochemical systems. Firstly, we constructed a glucose-fed fungus-bacteria microbial consortium including a fermenter (*Saccharomyces cerevisiae*) in which the ethanol pathway was knocked out and the lactic acid biosynthesis pathway was introduced into *Shewanella cerevisiae*, and an exoelectrogen (*S. oneidensis* MR-1). We optimized the co-culturing conditions of the microbial consortium to achieve an optimal coordination between carbon source metabolism of the fermenter and extracellular electron transfer of the exoelectrogen, such that lactate, the metabolic product of glucose by the recombinant *S. cerevisiae*, was continuously supplied to *S. oneidensis* in a constant level until glucose exhaustion. This metabolic coordination between the fermenter and the exoelectrogen enabled bioelectricity production in a glucose-fed MFC. Secondly, we used the design principle of "division-of-labor" to construct a three-species microbial consortium for power generation, consisting of recombinant *Escherichia coli*, *Bacillus subtilis* and *Shewanella oneidensis*. In this consortium, *E. coli* digested glucose to produce lactate as a carbon source, *B. subtilis* produced riboflavin as an electron shuttle, and *S. oneidensis* served as the exoelectrogen to generate electricity. The three species formed a cross-feeding microbial consortium, performing "better together" for power generation.

Our study provided new insights into the rational design of more efficient, stable, and robust synthetic microbial consortia applicable in bioenergy and environments.

References:
1. Yue Liu, Mingzhu Ding, Wei Ling, Yun Yang, Xiao Zhou, Bingzhi Li, Tao Chen, Yong Nie, Miaoxiao Wang, Boxuan Zeng, Xia Liwi, Hong Liu, Baode Sun, Heming Xu, Jiamei zhang, Yi Jiao, Yanan Hou, Hui Yang, Sijia Xiao, Qucheng Lin, Xinzi He, Wenjie Liao, Zeqi Jin, Yufei Xie, Bofeng Zhang, Tianyu Li, Xi Lu, Jiabei Li, Fan Zhang, Hao Song*, Yingjin Yuan*, "A three-species microbial consortium for power generation", Energy & Environmental Science, DOI: 10.1039/C6EE03705D (in press, 2017)

2. Tong Lin, Xue Bai, Yun Yang, Jingyu Wang, Yidan Hu, Bingzhi Li, Ying-Jin Yuan, Hao Song*, "Synthetic Saccharomyces cerevisiae-Shewanella oneidensis consortium enables glucose-fed microbial fuel cell", AIChE Journal, (in press, 2017)

496525 Screening Rhizobacteria Interactions Using a High-Throughput Microwell Array Platform

Abstract:

Plant root-associated microbial communities consist of thousands of unique species that function together to offer beneficial properties to the host. These benefits include nutrient uptake, plant hormone regulation, and pathogen protection. The structure and function of these communities is shaped by microbe-microbe interactions, most of which are unknown or poorly characterized because traditional co-culture methods can only examine a few interactions in a single assay. In addition, traditional methods typically probe interactions between colonies or bulk cell populations, overlooking single-cell interactions that often play a pivotal role in community function. To improve these characterization capabilities, we are developing a unique screening tool termed the â€œmicrobe arrayâ€, designed to discover unknown inter-cellular interactions that exist between plant pathogens and root microbiome members. By implementing previously developed techniques in microfabrication, the platform uses micro-scale wells to form thousands of unique combinations between small numbers of fluorescently-labeled plant pathogens and isolates from a root microbiome. Each well serves as a unique and independent co-culture site, and the growth of the pathogen in each well of the array can be monitored in parallel. This allows for identification of wells where abnormal pathogen growth occurs. We are currently developing materials and methods that enable extraction of cells out of individual wells of interest, allowing for the sequencing and discovery of microbiome members that promote or inhibit pathogen growth. Current studies are focused on screening interactions between *Agrobacterium tumefaciens*, the agent of Crown Gall Disease, and microbiome members from *Heliantus annuus* root. Identification of the critical species that inhibit plant pathogen growth and colonization will inform the development of constructed microbial communities for use in new biocontrol strategies. Successful development will also result in a new and powerful screening tool for rapidly uncovering microbe-microbe interactions that occur in any microbiome of interest.

498112 Engineering Modular Microbial Communities for Cellulose Utilization and Bioproduct Synthesis

Abstract:

Recent environmental concerns have increased the need to develop systems which can utilize waste materials to make chemicals. Cellulosic biomass, an abundant waste material, can be degraded into sugar monomers to serve as an inexpensive carbon source for microbial systems. Current approaches for biomass utilization rely on pretreatment processes and expensive purification schemes. To address these challenges, microbial communities can be engineered to synergistically perform complex tasks, by dividing the demands among multiple specialized modules, or functional units, to maximize the efficiency of desired product formation. Therefore, we aim to build microbial systems with interchangeable modules, with the first module dedicated to cellulose degradation and the second module specialized for bioproduct synthesis. To engineer synthetic communities for the production of biofuels and pharmaceuticals, we are building and testing mixed cultures of *Bacillus megaterium* and *Escherichia coli* strains. We will describe our recent progress engineering these two modules and assembling them for activity.

Complete degradation of cellulose requires a combination of three cellulase activities â€" an endoglucanase, a cellobiohydrolase, and a beta-glucosidase. We have developed a cellulose degradation module consisting of *Bacillus megaterium* strains for the secretion of cellulases. A small library of signal peptides (SPs) was selected to tag each cellulase, and thereby used to trigger protein secretion in *B. megaterium*. Cellulase activity against amorphous cellulose was confirmed through a series of bioassays, and the most active signal peptide constructs were identified for each cellulase. Expression optimization for each strain focused on testing secretion and activity under different expression temperatures, induction time points, expression lengths, and media conditions. Activity of the optimized cellulase secretion strains was characterized individually and in tandem to assess synergistic cellulolytic activity. We have also demonstrated that the secreted cellulase activity can be used to degrade cellulose into glucose monomers to support growth of an *E. coli* production module. We will describe our progress investigating the effect of different inoculation ratios, expression conditions, and community compositions for the cellulose degradation module to determine optimized cellulase secretion and activity conditions for the system. Further system development has also targeted the second module, with the introduction of a violacein-producing *E. coli* strain. The cellulose degradation module can be integrated with many different bioproduct synthesis modules to build complex systems for the synthesis of high value products from cellulose.

498559 Symbiotic Microbial Communities for Cleaning Agricultural Waters and Bioenergy Production

Abstract:

With increase in intensive crop and livestock production, the excess application of P as fertilizer and manure contributes to the build-up of soil P levels causing eutrophication [1, 2]. Currently a curative solution to combat water pollution by recuperating N and P, recycling the P for food production by closing the nutrient loop to effectively manage dwindling rock P reserve are the major challenges. The most common phosphorus removal techniques for municipal and industrial wastewater are biological treatment and chemical precipitation. Bio-valorization offers an excellent alternative to chemical treatment technologies as the biomass produced can be a source of protein and other valuable bio-chemicals and bio-fuels in this low cost technology. In this present work, a novel symbiotic microbial community called mycoalgae biofilm [3] was developed using the isolated strains of polyphosphate accumulating fungi and microalgae for cleaning agricultural waters and bioenergy production. We want to apply this platform for the efficient recovery and recycling of nutrients in waste stream. The effect of various process conditions on the nutrient recovery efficiency and the changes in the biofilm biomass composition was studied. Since the biofilm will be used for nutrient recovery from an array of samples, a synthetic medium for the cell growth was used as the medium for the biofilm development to have different initial conditions. Depending on the nature of the wastewater and its composition, and the fungi-microalgae combination intended to use for bio-remediation, the developed mycoalgae biofilm can be a sustainable microbial process for food, feed or energy source. The composite biofilm shows promising results in wastewater treatment especially in removing the N, P, suspended solids and other organics from nutrient polluted water in a single-step reactor with easy biomass recovery and high efficiency, when compared to the individual pure cell cultures. The biofilm attaches to the polymer matrix, leaving behind the clean water for recirculation, and the composite biomass for nutrient recycle. The commercial expansion of this biofilm system could be a step ahead in having integrated and sustainable water treatment systems.

References

1. E.K. Wilson, Danger from microcystins in Toledo water unclear, Chem. Eng. News, 92 (32) (2014), p. 9
2. Ye, Y., Gan, J., & Hu, B. (2015). Screening of phosphorus-accumulating fungi and their potential for Phosphorus removal from waste streams. Applied Biochemistry and Biotechnology, 177(5), 1127-1136.
3. Aravindan Rajendran and Bo Hu, Mycoalgae biofilm: development of a novel platform technology using algae and fungal cultures, Biotechnology for Biofuels, 20169:112, DOI: 10.1186/s13068-016-0533-y

AUTHOR INDEX